Chemistry Laboratory Notebook

This laboratory notebook belongs to

Name _____

Address _____

Telephone number _____

Course _____ Section _____

Laboratory instructor's name _____

Laboratory partner's name _____

Equipment drawer or locker number _____

Date: from _____ to _____

Telephone Numbers of Emergency Services

Campus police _____

Campus fire _____

Health center _____

Poison control center _____

Suggestions for Using This Laboratory Notebook

1. Fill in your name, address, and telephone number on the first page so your notebook can be returned if lost.

2. Create a table of contents for your notebook as you use it during the course. Record the titles of the experiments next to the page numbers on which your write-up can be found.

3. Notebook entries are a **permanent record** of your laboratory work. Never remove the original (white) pages. The yellow (copy) pages are perforated for easy removal, if you are required to turn in a copy of your work.

4. Always **write in ink** using a firm-tipped pen, such as a ballpoint. Use enough pressure to ensure that a clear, dark copy is made.

5. If you make a mistake, draw a single line through the error and write the correct entry nearby.

6. Provide the information asked for at the top of each notebook page.

7. Use the left column to record data and calculations. In the right column write a report of your work. Include all of the elements required by your laboratory instructor, such as the purpose of the experiment, the procedure used, observations made, and conclusions drawn.

Table of Contents

51 _____

52 _____

53 _____

54 _____

55 _____

56 _____

57 _____

58 _____

59 _____

60 _____

61 _____

62 _____

63 _____

64 _____

65 _____

66 _____

67 _____

68 _____

69 _____

70 _____

71 _____

72 _____

73 _____

74 _____

75 _____

76 _____

77 _____

78 _____

79 _____

80 _____

81 _____

82 _____

83 _____

84 _____

85 _____

86 _____

87 _____

88 _____

89 _____

90 _____

91 _____

92 _____

93 _____

94 _____

95 _____

96 _____

97 _____

98 _____

99 _____

100 _____

Safety Contract[*]

Whenever I am in an area where laboratory reagents are being used, I agree to abide by the following rules:

1. Wear safety goggles.

2. Wear proper clothing.

3. Use good housekeeping practices.

4. Do only authorized experiments, and work only when the laboratory instructor or another qualified person is present.

5. Treat laboratory reagents as if they are poisonous and corrosive.

6. Dispense reagents carefully. Dispose of laboratory reagents as directed.

7. Not eat, drink, use tobacco, or apply cosmetics in the laboratory.

8. Report all incidents to the laboratory instructor.

9. Be familiar with the location and use of all safety equipment.

10. Become familiar with each laboratory assignment before coming to the laboratory.

11. Anticipate the common hazards that may be encountered in laboratory.

12. Become familiar with actions to be taken in the event of incidents in the laboratory.

student signature _____ date _____

laboratory instructor _____ date _____

In the space below, give any health information, such as pregnancy or other circumstance, that might help the laboratory instructor provide a safer environment for you, or that could aid the laboratory instructor in responding to an incident involving you in the laboratory.

1. I do/do not (circle one) expect to wear contact lenses during laboratory work. [Note: Goggles must still be worn when contact lenses are worn.]

2. List any known allergies to medication or other chemicals.

*From Rapp, M.W. *Practicing Safety in the Organic Chemistry Laboratory*; Chemical Education Resources: Palmyra, PA, 1997

Experiment title and number		Date	
Name	Course	Section	Lab partner

Caution: Place fold-in flap under yellow sheet before writing, to protect the pages that follow.

Experiment title and number		Date	
Name		Course	Section	Lab partner

Caution: Place fold-in flap under yellow sheet before writing, to protect the pages that follow.

Experiment title and number

Date

Name

Course

Section

Lab partner

Experiment title and number

Date

Name

Course

Section

Lab partner

Caution: Place fold-in flap under yellow sheet before writing, to protect the pages that follow.

Experiment title and number

Date

Name

Course

Section

Lab partner

Caution: Place fold-in flap under yellow sheet before writing, to protect the pages that follow.

Experiment title and number

Date

Name

Course

Section

Lab partner

Caution: Place fold-in flap under yellow sheet before writing, to protect the pages that follow.

Experiment title and number

Date

Name

Course

Section

Lab partner

Experiment title and number

Date

04

Name

Course

Section

Lab partner

Caution: Place fold-in flap under yellow sheet before writing, to protect the pages that follow.

Experiment title and number

Date

Name

Course

Section

Lab partner

Experiment title and number

Date

Caution: Place fold-in flap under yellow sheet before writing, to protect the pages that follow.

Name

Course

Section

Lab partner

Caution: Place fold-in flap under yellow sheet before writing, to protect the pages that follow.

Experiment title and number		Date	
Name	Course	Section	Lab partner

Caution: Place fold-in flap under yellow sheet before writing, to protect the pages that follow.

Experiment title and number

Date

06

Name

Course

Section

Lab partner

Experiment title and number

Date

07

Name

Course

Section

Lab partner

Caution: Place fold-in flap under yellow sheet before writing, to protect the pages that follow.

Experiment title and number		Date

Name	Course	Section	Lab partner

Caution: Place fold-in flap under yellow sheet before writing, to protect the pages that follow.

Experiment title and number

Date

Name

Course

Section

Lab partner

Experiment title and number

Date

Name

Course

Section

Lab partner

Caution: Place fold-in flap under yellow sheet before writing, to protect the pages that follow.

Name

Course

Section

Lab partner

Caution: Place fold-in flap under yellow sheet before writing, to protect the pages that follow.

Experiment title and number

Date

Name

Course

Section

Lab partner

Date

Caution: Place fold-in flap under yellow sheet before writing, to protect the pages that follow.

Experiment title and number

Date

Name

Course

Section

Lab partner

Experiment title and number		Date	

Name	Course	Section	Lab partner

Caution: Place fold-in flap under yellow sheet before writing, to protect the pages that follow.

Experiment title and number

Date

Name

Course

Section

Lab partner

Caution: Place fold-in flap under yellow sheet before writing, to protect the pages that follow.

Experiment title and number

Date

Name

Course

Section

Lab partner

Caution: Place fold-in flap under yellow sheet before writing, to protect the pages that follow.

Experiment title and number		Date	
Name	Course	Section	Lab partner

Caution: Place fold-in flap under yellow sheet before writing, to protect the pages that follow.

Experiment title and number		Date

Name		Course	Section	Lab partner

Caution: Place fold-in flap under yellow sheet before writing, to protect the pages that follow.

Name

Course

Section

Lab partner

Caution: Place fold-in flap under yellow sheet before writing, to protect the pages that follow.

Experiment title and number

Date

Name

Course

Section

Lab partner

Name

Course

Section

Lab partner

Caution: Place fold-in flap under yellow sheet before writing, to protect the pages that follow.

Experiment title and number

Date

Name

Course

Section

Lab partner

Experiment title and number		Date	**15**
Name	Course	Section	Lab partner

Experiment title and number

Date

Name

Course

Section

Lab partner

Experiment title and number

Date

Name

Course

Section

Lab partner

Caution: Place fold-in flap under yellow sheet before writing, to protect the pages that follow.

Experiment title and number

Date

Name

Course

Section

Lab partner

Caution: Place fold-in flap under yellow sheet before writing, to protect the pages that follow.

Experiment title and number		Date

Name	Course	Section	Lab partner

Experiment title and number		Date

Caution: Place fold-in flap under yellow sheet before writing, to protect the pages that follow.

Name

Course

Section

Lab partner

Caution: Place fold-in flap under yellow sheet before writing, to protect the pages that follow.

Experiment title and number

Date

Name

Course

Section

Lab partner

Experiment title and number

Date

Name

Course

Section

Lab partner

Experiment title and number

Date

Name

Course

Section

Lab partner

Caution: Place fold-in flap under yellow sheet before writing, to protect the pages that follow.

Experiment title and number

Date

20

Name

Course

Section

Lab partner

Caution: Place fold-in flap under yellow sheet before writing, to protect the pages that follow.

Name

Course

Section

Lab partner

Caution: Place fold-in flap under yellow sheet before writing, to protect the pages that follow.

Experiment title and number

Date

Name

Course

Section

Lab partner

Experiment title and number

Date

Caution: Place fold-in flap under yellow sheet before writing, to protect the pages that follow.

Experiment title and number		Date	
Name	Course	Section	Lab partner

Experiment title and number		Date	
Name	Course	Section	Lab partner

Experiment title and number		Date	
Name	Course	Section	Lab partner

Caution: Place fold-in flap under yellow sheet before writing, to protect the pages that follow.

Experiment title and number		Date
Name	Course	Section	Lab partner

Experiment title and number		Date	

Name	Course	Section	Lab partner

Caution: Place fold-in flap under yellow sheet before writing, to protect the pages that follow.

Experiment title and number

Date

Name

Course

Section

Lab partner

Experiment title and number

Date

Name

Course

Section

Lab partner

Caution: Place fold-in flap under yellow sheet before writing, to protect the pages that follow.

Experiment title and number

Date

Name

Course

Section

Lab partner

Name

Course

Section

Lab partner

Experiment title and number

Date

Name

Course

Section

Lab partner

Experiment title and number		Date	
Name	Course	Section	Lab partner

Caution: Place fold-in flap under yellow sheet before writing, to protect the pages that follow.

Experiment title and number

Date

Name

Course

Section

Lab partner

Experiment title and number

Date

Name

Course

Section

Lab partner

Caution: Place fold-in flap under yellow sheet before writing, to protect the pages that follow.

Experiment title and number

Date

Name

Course

Section

Lab partner

Experiment title and number

Date

Caution: Place fold-in flap under yellow sheet before writing, to protect the pages that follow.

Name

Course

Section

Lab partner

Caution: Place fold-in flap under yellow sheet before writing, to protect the pages that follow.

Experiment title and number

Date

Name

Course   Section   Lab partner

Experiment title and number

Date

Experiment title and number		Date
Name	Course	Section	Lab partner

Name   Course   Section   Lab partner

Caution: Place fold-in flap under yellow sheet before writing, to protect the pages that follow.

Experiment title and number

Date

Name

Course

Section

Lab partner

Caution: Place fold-in flap under yellow sheet before writing, to protect the pages that follow.

Experiment title and number		Date
Name	Course	Section	Lab partner

Caution: Place fold-in flap under yellow sheet before writing, to protect the pages that follow.

Experiment title and number

Date

Name

Course

Section

Lab partner

Experiment title and number

Date

Caution: Place fold-in flap under yellow sheet before writing, to protect the pages that follow.

Experiment title and number

Date

Name

Course

Section

Lab partner

Experiment title and number		Date	
Name	Course	Section	Lab partner

Caution: Place fold-in flap under yellow sheet before writing, to protect the pages that follow.

Experiment title and number

Date

Name

Course

Section

Lab partner

Name   Course   Section   Lab partner

Caution: Place fold-in flap under yellow sheet before writing, to protect the pages that follow.

Experiment title and number

Date

Name

Course

Section

Lab partner

Experiment title and number

Date

Name

Course

Section

Lab partner

Experiment title and number

Date

Name

Course

Section

Lab partner

Experiment title and number

Date

Name

Course

Section

Lab partner

Experiment title and number

Date

Caution: Place fold-in flap under yellow sheet before writing, to protect the pages that follow.

Experiment title and number		Date
Name	Course	Section	Lab partner

Name

Course

Section

Lab partner

Caution: Place fold-in flap under yellow sheet before writing, to protect the pages that follow.

Experiment title and number

Date

Name

Course

Section

Lab partner

Experiment title and number

Date

Caution: Place fold-in flap under yellow sheet before writing, to protect the pages that follow.

Experiment title and number		Date	
Name	Course	Section	Lab partner

Name   Course   Section   Lab partner

Caution: Place fold-in flap under yellow sheet before writing, to protect the pages that follow.

Experiment title and number		Date

Name	Course	Section	Lab partner

Experiment title and number		Date

Caution: Place fold-in flap under yellow sheet before writing, to protect the pages that follow.

Name

Course

Section

Lab partner

Experiment title and number		Date	
Name	Course	Section	Lab partner

Caution: Place fold-in flap under yellow sheet before writing, to protect the pages that follow.

Name

Course

Section

Lab partner

Caution: Place fold-in flap under yellow sheet before writing, to protect the pages that follow.

Experiment title and number

Date

Name

Course

Section

Lab partner

Experiment title and number

Date

Name

Course

Section

Lab partner

Caution: Place fold-in flap under yellow sheet before writing, to protect the pages that follow.

Experiment title and number

Date

Name

Course

Section

Lab partner

Caution: Place fold-in flap under yellow sheet before writing, to protect the pages that follow.

Experiment title and number

Date

Name

Course

Section

Lab partner

Experiment title and number

Date

Name

Course

Section

Lab partner

Name

Course

Section

Lab partner

Caution: Place fold-in flap under yellow sheet before writing, to protect the pages that follow.

Experiment title and number

Date

44

Name

Course

Section

Lab partner

Experiment title and number		Date

Name	Course	Section	Lab partner

Caution: Place fold-in flap under yellow sheet before writing, to protect the pages that follow.

Experiment title and number

Date

Name

Course

Section

Lab partner

Experiment title and number

Date

Name

Course

Section

Lab partner

Caution: Place fold-in flap under yellow sheet before writing, to protect the pages that follow.

Experiment title and number

Date

Name

Course

Section

Lab partner

Caution: Place fold-in flap under yellow sheet before writing, to protect the pages that follow.

Experiment title and number		Date	
Name	Course	Section	Lab partner

Experiment title and number

Date

Name

Course

Section

Lab partner

Experiment title and number		Date	47

Name	Course	Section	Lab partner

Experiment title and number		Date	47

Experiment title and number

Date

Name

Course

Section

Lab partner

Caution: Place fold-in flap under yellow sheet before writing, to protect the pages that follow.

Caution: Place fold-in flap under yellow sheet before writing, to protect the pages that follow.

Experiment title and number

Date

Name

Course

Section

Lab partner

Caution: Place fold-in flap under yellow sheet before writing, to protect the pages that follow.

Experiment title and number   Date   49

Name   Course   Section   Lab partner

Caution: Place fold-in flap under yellow sheet before writing, to protect the pages that follow.

Experiment title and number

Date

Name

Course

Section

Lab partner

Caution: Place fold-in flap under yellow sheet before writing, to protect the pages that follow.

Experiment title and number

Date

50

Name

Course

Section

Lab partner

Caution: Place fold-in flap under yellow sheet before writing, to protect the pages that follow.

Experiment title and number

Date

Name

Course

Section

Lab partner

Caution: Place fold-in flap under yellow sheet before writing, to protect the pages that follow.

Experiment title and number

Date

Name

Course

Section

Lab partner

Experiment title and number

Date

Caution: Place fold-in flap under yellow sheet before writing, to protect the pages that follow.

Experiment title and number		Date	
Name	Course	Section	Lab partner

Caution: Place fold-in flap under yellow sheet before writing, to protect the pages that follow.

Experiment title and number

Date

Name

Course

Section

Lab partner

Caution: Place fold-in flap under yellow sheet before writing, to protect the pages that follow.

Experiment title and number   Date

Name   Course   Section   Lab partner

Experiment title and number

Date

Name

Course

Section

Lab partner

Experiment title and number

Date

Caution: Place fold-in flap under yellow sheet before writing, to protect the pages that follow.

Experiment title and number

Date

Name

Course

Section

Lab partner

Experiment title and number

Date

Name

Course

Section

Lab partner

Experiment title and number

Date

Caution: Place fold-in flap under yellow sheet before writing, to protect the pages that follow.

Name

Course

Section

Lab partner

Caution: Place fold-in flap under yellow sheet before writing, to protect the pages that follow.

Name

Course

Section

Lab partner

Caution: Place fold-in flap under yellow sheet before writing, to protect the pages that follow.

55

Name

Course

Section

Lab partner

Experiment title and number

Date

Name

Course

Section

Lab partner

Caution: Place fold-in flap under yellow sheet before writing, to protect the pages that follow.

Experiment title and number		Date	
Name	Course	Section	Lab partner

Experiment title and number

Date

Name

Course

Section

Lab partner

Experiment title and number

Date

Caution: Place fold-in flap under yellow sheet before writing, to protect the pages that follow.

Name

Course

Section

Lab partner

Name

Course

Section

Lab partner

Caution: Place fold-in flap under yellow sheet before writing, to protect the pages that follow.

Experiment title and number

Date

Name

Course

Section

Lab partner

Name

Course

Section

Lab partner

Caution: Place fold-in flap under yellow sheet before writing, to protect the pages that follow.

Experiment title and number

Date

Name

Course

Section

Lab partner

Name

Course

Section

Lab partner

Caution: Place fold-in flap under yellow sheet before writing, to protect the pages that follow.

Experiment title and number

Date

Name

Course

Section

Lab partner

Caution: Place fold-in flap under yellow sheet before writing, to protect the pages that follow.

Experiment title and number		Date
Name	Course	Section	Lab partner

Caution: Place fold-in flap under yellow sheet before writing, to protect the pages that follow.

Name

Course

Section

Lab partner

Caution: Place fold-in flap under yellow sheet before writing, to protect the pages that follow.

Experiment title and number

Date

Name

Course

Section

Lab partner

Experiment title and number

Date

Name

Course

Section

Lab partner

Experiment title and number

Date

Name

Course

Section

Lab partner

Caution: Place fold-in flap under yellow sheet before writing, to protect the pages that follow.

Experiment title and number

Date

Name

Course

Section

Lab partner

Caution: Place fold-in flap under yellow sheet before writing, to protect the pages that follow.

Experiment title and number

Date

Name

Course

Section

Lab partner

Caution: Place fold-in flap under yellow sheet before writing, to protect the pages that follow.

65

Name

Course

Section

Lab partner

Caution: Place fold-in flap under yellow sheet before writing, to protect the pages that follow.

Name

Course

Section

Lab partner

Experiment title and number

Date

Name

Course

Section

Lab partner

Caution: Place fold-in flap under yellow sheet before writing, to protect the pages that follow.

Name

Course

Section

Lab partner

Caution: Place fold-in flap under yellow sheet before writing, to protect the pages that follow.

Experiment title and number

Date

Name

Course

Section

Lab partner

Experiment title and number

Date

67

Caution: Place fold-in flap under yellow sheet before writing, to protect the pages that follow.

Experiment title and number

Date

Name

Course

Section

Lab partner

Caution: Place fold-in flap under yellow sheet before writing, to protect the pages that follow.

Name

Course

Section

Lab partner

Caution: Place fold-in flap under yellow sheet before writing, to protect the pages that follow.

Experiment title and number		Date	
Name	Course	Section	Lab partner

Experiment title and number

Date

Name

Course

Section

Lab partner

Experiment title and number

Date

Caution: Place fold-in flap under yellow sheet before writing, to protect the pages that follow.

Experiment title and number

Date

Name

Course

Section

Lab partner

Experiment title and number

Date

70

Name
Course
Section
Lab partner

Experiment title and number

Date

70

Caution: Place fold-in flap under yellow sheet before writing, to protect the pages that follow.

Caution: Place fold-in flap under yellow sheet before writing, to protect the pages that follow.

Experiment title and number

Date

Name

Course

Section

Lab partner

Experiment title and number

Date

Caution: Place fold-in flap under yellow sheet before writing, to protect the pages that follow.

Experiment title and number

Date

Name

Course

Section

Lab partner

Caution: Place fold-in flap under yellow sheet before writing, to protect the pages that follow.

Experiment title and number

Date

Name

Course

Section

Lab partner

Caution: Place fold-in flap under yellow sheet before writing, to protect the pages that follow.

Experiment title and number

Date

Name

Course

Section

Lab partner

Experiment title and number

Date

Caution: Place fold-in flap under yellow sheet before writing, to protect the pages that follow.

Name

Course

Section

Lab partner

Caution: Place fold-in flap under yellow sheet before writing, to protect the pages that follow.

Experiment title and number

Date

Name

Course

Section

Lab partner

Caution: Place fold-in flap under yellow sheet before writing, to protect the pages that follow.

Experiment title and number	Date

Name	Course	Section	Lab partner

Experiment title and number	Date

Caution: Place fold-in flap under yellow sheet before writing, to protect the pages that follow.

Experiment title and number   Date

Name   Course   Section   Lab partner

Caution: Place fold-in flap under yellow sheet before writing, to protect the pages that follow.

Experiment title and number			Date	
Name		Course	Section	Lab partner

Experiment title and number

Date

Name

Course

Section

Lab partner

Experiment title and number

Date

Name

Course

Section

Lab partner

Experiment title and number

Date

Caution: Place fold-in flap under yellow sheet before writing, to protect the pages that follow.

Experiment title and number

Date

Name

Course

Section

Lab partner

Caution: Place fold-in flap under yellow sheet before writing, to protect the pages that follow.

Experiment title and number

Date

Name

Course

Section

Lab partner

Caution: Place fold-in flap under yellow sheet before writing, to protect the pages that follow.

Experiment title and number

Date

Name

Course

Section

Lab partner

Experiment title and number		Date	
Name	Course	Section	Lab partner

Experiment title and number

Date

Name

Course

Section

Lab partner

Caution: Place fold-in flap under yellow sheet before writing, to protect the pages that follow.

Name

Course

Section

Lab partner

Caution: Place fold-in flap under yellow sheet before writing, to protect the pages that follow.

Name

Course

Section

Lab partner

Experiment title and number

Date

Name

Course

Section

Lab partner

Caution: Place fold-in flap under yellow sheet before writing, to protect the pages that follow.

Caution: Place fold-in flap under yellow sheet before writing, to protect the pages that follow.

Experiment title and number

Date

Name

Course

Section

Lab partner

Name

Course

Section

Lab partner

Caution: Place fold-in flap under yellow sheet before writing, to protect the pages that follow.

Experiment title and number

Date

Name

Course

Section

Lab partner

Caution: Place fold-in flap under yellow sheet before writing, to protect the pages that follow.

Name

Course

Section

Lab partner

Caution: Place fold-in flap under yellow sheet before writing, to protect the pages that follow.

Experiment title and number

Date

Name

Course

Section

Lab partner

Name

Course

Section

Lab partner

Caution: Place fold-in flap under yellow sheet before writing, to protect the pages that follow.

Experiment title and number

Date

Name

Course

Section

Lab partner

Caution: Place fold-in flap under yellow sheet before writing, to protect the pages that follow.

Experiment title and number

Date

Name

Course

Section

Lab partner

Caution: Place fold-in flap under yellow sheet before writing, to protect the pages that follow.

Experiment title and number

Date

Name

Course

Section

Lab partner

Caution: Place fold-in flap under yellow sheet before writing, to protect the pages that follow.

Experiment title and number   Date

Name   Course   Section   Lab partner

Name

Course

Section

Lab partner

Caution: Place fold-in flap under yellow sheet before writing, to protect the pages that follow.

Experiment title and number

Date

Name

Course

Section

Lab partner

Caution: Place fold-in flap under yellow sheet before writing, to protect the pages that follow.

Name

Course

Section

Lab partner

Caution: Place fold-in flap under yellow sheet before writing, to protect the pages that follow.

Experiment title and number

Date

Name

Course

Section

Lab partner

Caution: Place fold-in flap under yellow sheet before writing, to protect the pages that follow.

Experiment title and number			Date	
Name		Course	Section	Lab partner

Caution: Place fold-in flap under yellow sheet before writing, to protect the pages that follow.

Experiment title and number

Date

Name

Course

Section

Lab partner

Caution: Place fold-in flap under yellow sheet before writing, to protect the pages that follow.

Name

Course

Section

Lab partner

Caution: Place fold-in flap under yellow sheet before writing, to protect the pages that follow.

Experiment title and number   Date   91

Name   Course   Section   Lab partner

Caution: Place fold-in flap under yellow sheet before writing, to protect the pages that follow.

Experiment title and number

Date

Name

Course

Section

Lab partner

Experiment title and number

Date

Caution: Place fold-in flap under yellow sheet before writing, to protect the pages that follow.

Experiment title and number

Date

Name

Course

Section

Lab partner

Experiment title and number

Date

Name

Course

Section

Lab partner

Caution: Place fold-in flap under yellow sheet before writing, to protect the pages that follow.

Experiment title and number

Date

Name

Course

Section

Lab partner

Caution: Place fold-in flap under yellow sheet before writing, to protect the pages that follow.

Name		Course	Section	Lab partner

Caution: Place fold-in flap under yellow sheet before writing, to protect the pages that follow.

Experiment title and number

Date

Name

Course

Section

Lab partner

Caution: Place fold-in flap under yellow sheet before writing, to protect the pages that follow.

Experiment title and number		Date	

Name	Course	Section	Lab partner

Caution: Place fold-in flap under yellow sheet before writing, to protect the pages that follow.

Experiment title and number

Date

Name

Course

Section

Lab partner

Caution: Place fold-in flap under yellow sheet before writing, to protect the pages that follow.

Experiment title and number

Date

Name

Course

Section

Lab partner

Caution: Place fold-in flap under yellow sheet before writing, to protect the pages that follow.

Experiment title and number

Date

Name

Course

Section

Lab partner

Experiment title and number

Date

97

Name

Course

Section

Lab partner

Experiment title and number

Date

97

Caution: Place fold-in flap under yellow sheet before writing, to protect the pages that follow.

Name

Course

Section

Lab partner

Caution: Place fold-in flap under yellow sheet before writing, to protect the pages that follow.

Experiment title and number

Date

Name

Course

Section

Lab partner

Caution: Place fold-in flap under yellow sheet before writing, to protect the pages that follow.

Experiment title and number

Date

Name

Course

Section

Lab partner

Caution: Place fold-in flap under yellow sheet before writing, to protect the pages that follow.

| Experiment title and number | | | | | | Date | | | | 99 |

| Name | | | | Course | | Section | | Lab partner | | |

Caution: Place fold-in flap under yellow sheet before writing, to protect the pages that follow.

Experiment title and number		Date	

Name	Course	Section	Lab partner

Caution: Place fold-in flap under yellow sheet before writing, to protect the pages that follow.

Experiment title and number

Date

Name

Course

Section

Lab partner

Caution: Place fold-in flap under yellow sheet before writing, to protect the pages that follow.